Generation Us

The Challenge of Global Warming

ANDREW WEAVER

RAVEN BOOKS
an imprint of
ORCA BOOK PUBLISHERS

Library and Archives Canada Cataloguing in Publication

Weaver, Andrew, 1961-
Generation us : the challenge of global warming / written by Andrew Weaver.

Issued also in electronic format.
ISBN 978-1-55469-804-2

1. Readers (Adult). 2. High interest-low vocabulary books. 3. English
language--Textbooks for second language learners. 4. Global warming.

I. Title. II. Series: Rapid reads
PE1126.A4W32 2011   428.6'2   C2011-900332-5

First published in the United States, 2011
Library of Congress Control Number: 2010943311

Summary: The issues surrounding global warming are explained and solutions
offered by one of the world's leading experts in the field.

*Orca Book Publishers is dedicated to preserving the environment and has printed
this book on paper certified by the Forest Stewardship Council®.*

Orca Book Publishers gratefully acknowledges the support for
its publishing programs provided by the following agencies:
the Government of Canada through the Canada Book Fund and the Canada
Council for the Arts, and the Province of British Columbia through the BC
Arts Council and the Book Publishing Tax Credit.

Design by Teresa Bubela
Cover photograph by Getty Images

ORCA BOOK PUBLISHERS
PO Box 5626, Stn. B
Victoria, BC Canada
V8R 6S4

ORCA BOOK PUBLISHERS
PO Box 468
Custer, WA USA
98240-0468

www.orcabook.com
Printed and bound in Canada.

14 13 12 11 • 4 3 2 1

To the youth of today,
the decision makers of tomorrow.

The Nature of Science and the Science of Nature

These days it seems hardly a day goes by that we don't hear mention of the topic of **global warming**. (*Words in bold can be found in the Glossary of Terms on page 118.*) It's on the news and in the newspapers. It's even in our daily conversation with friends. Every time there is a flood, a drought, a heavy snowfall or strong winds, someone somewhere blames it on global warming. And someone else tells us that there's nothing to worry about—these are just freak weather events. So who is right? Is global warming a natural phenomenon? Can we blame individual weather events

1

on global warming? To answer these questions, we first have to clarify two issues: the meaning of the **scientific method** and the difference between **weather** and **climate**.

THE SCIENTIFIC METHOD

The word *science* comes from the Latin word *scientia*, meaning knowledge or skill. Nowadays, we use the word to describe our knowledge of either the natural world (natural sciences) or human society (social sciences). Subjects like physics, chemistry and biology are examples of the natural sciences, whereas economics, sociology and psychology are social sciences. In both cases, new knowledge is acquired by following a process known as the scientific method.

In the scientific method, a natural or social scientist starts off by identifying a phenomenon that he or she wants to try to understand. For example, a biologist might want to know why cedar trees grow better in one valley than in another. The scientist then observes the phenomenon by collecting data that help describe it as completely as possible. This is called the observational stage. In the case of the biologist, he or she might collect tree and soil samples for analysis back in the laboratory. He or she might also try to find out what nearby weather data are available.

Next, the scientist will make an educated guess as to why the phenomenon appears the way it does. This is known as the hypothesis stage. For example, after examining the soils from both valleys and finding out that one contains fewer nutrients (natural fertilizer), the biologist might suppose that this observation explains the difference in how the cedar trees grow. In the third stage, the scientist uses the educated guess (or **hypothesis**) to predict what will happen to the phenomenon in different situations. This is called the prediction stage. Our biologist might predict that the struggling cedar trees would grow better if nutrients (fertilizer) were added to their soils.

In the final **experimental** stage, experiments are designed to test the **predictions**. If the predictions are correct, the hypothesis stands—at least for now. If the predictions fail, the new experimental observations would have to be recorded, the hypothesis or explanation would have to be modified, and the whole process would repeat itself.

So if, after adding fertilizer, our biologist finds that there is no effect on the cedar trees, he or she will have to come up with a new hypothesis and new experiments. The biologist might have noted in the observational stage that rainfall was different in the two valleys and that the thriving cedar trees received more rain than the other trees. He or she could then conduct an experiment that involved watering the trees in the drier valley. If they started to thrive, the biologist would have an explanation for the differences in the growth pattern of the trees in the two valleys.

In summary, science seeks to develop the understanding of a particular phenomenon that explains all known observations of it. In this book the phenomenon that we seek to explain is the Earth's warming over the last 130 years. This is the phenomenon that we know as global warming.

CLIMATE IS DIFFERENT FROM WEATHER

When you get up in the morning and look out the window to decide what clothes you are going to wear, you are concerned about the weather. In January, you are thinking about the climate as you pack appropriate clothes in your suitcase before heading to Hawaii for a winter holiday. Put another way, climate is what you expect; weather is what you get.

We expect the average temperature at JFK Airport in New York to be 32°F (0°C) in January, but on any given January day it may be much warmer or colder. Climate refers to the likelihood of occurrence of any given weather event rather than the weather event itself. This is important since it means that scientists can never blame an individual hurricane, rainfall, heat wave or drought on global warming. Rather, they can only say that such events may become more or less likely in a particular region. And they can assign a probability to how much more or less likely they will be.

Understanding the distinction between weather and climate is an important first step in understanding how scientists are able to make future projections about global warming. Climate scientists do not make long-range weather predictions. They instead try to estimate the change in average conditions and the likelihood of occurrence of any particular weather event. For example, making a prediction that next Saturday in Washington, DC, it will be cloudy with sunny breaks, with a high of 68°F (20°C) and a low of 59°F (15°C), is a weather forecast. Predicting that average summer temperatures over eastern North America will warm by 6°F (3°C) by the end of this century and that, at the same time, total winter precipitation will increase by about 11% is a climate prediction. Predicting that summer will be warmer than winter is a pretty safe bet. But by doing so, you are making a climate prediction. Understanding global warming involves understanding what affects the amount of energy the Earth receives from the sun and the amount of energy the Earth emits back to space.

DEALING WITH UNCERTAINTY

We constantly make decisions in the face of uncertainty. We eat food without knowing exactly where it came from or who handled it on its way to the supermarket shelf. We drive to work without knowing who else will be on the roads. We go to school not knowing exactly what will happen in class. We are constantly assessing the risk versus reward of our actions in the face of uncertainty. For example, in 2007 there were 42,031 deaths from motor-vehicle accidents in the United States. This means that if you live in the United States you have a 1 in 58 chance of dying from a motor-vehicle accident. That's pretty steep odds, but we still drive in cars.

All science is uncertain. It's important to remember that there is no such thing as *proof* in any field of science. Scientific laws and theories only explain all known observations of a particular phenomenon. If an existing theory or law can't explain a new observation, the theory or law has to be modified. Perhaps the most famous example of this happened in 1915 when Albert Einstein developed his

General Theory of Relativity. This modified the Theory of Gravity first developed by Isaac Newton in the seventeenth century.

AN OUTLINE OF WHAT FOLLOWS

In the next section, I describe observations related to the phenomenon of global warming. I provide the explanations for these observations that scientists have developed using the scientific method. And I show how these explanations have led scientists to make projections of what some of the effects of global warming might be.

In Part Two, I take the knowledge gained from Part One to outline some of the reasons why we might be concerned about global warming. Here it will be important to understand scientific uncertainty as well as how we assess risk in the face of this uncertainty. Finally, in Part Three, I offer some solutions that we can collectively use to combat global warming.

After you read this book, I hope you will be convinced that global warming is occurring. There is also overwhelming scientific evidence that it is largely caused by humans. My goal is to detail some of the reasons why we might care about the consequences of global warming. I will point out that there are readily available technological and policy solutions to the problem. Since everyone is part of the problem, everyone is also part of the solution. But whether we wish to deal with global warming or not seems to depend on our answer to the following question: What, if anything, does the present generation owe to future generations?

PART ONE

What Is the Problem?

Global warming describes the average warming of the Earth's surface temperatures as a consequence of human activity. Over the last 130 years, the Earth's surface air temperature has been warming by about 0.11°F (0.06°C) per decade (see Figure 1.1). However, the rate of warming has more than doubled, to 0.28°F (0.16°C) per decade, over the last thirty years. In fact, the warmest decade since humans have been compiling global temperature measurements is the 2000s. The second-warmest decade is the 1990s. The ten warmest years in descending order are: 2010, 2005, 1998, 2003, 2002, 2009, 2006, 2007, 2004 and 2001. The eleventh warmest year is 2008.

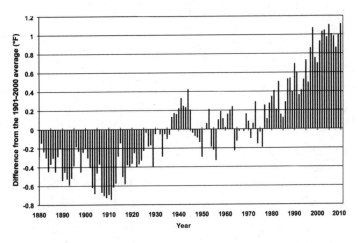

FIGURE 1.1: *Annually and globally averaged surface air temperature difference from the 1901–2000 average, expressed in °F. (Data collected by the US National Oceanic and Atmospheric Administration [NOAA].)*

Global warming does not mean that the Earth will warm by the same amount and at the same rate everywhere. Some regions warm faster than others. For example, the Arctic is expected to warm 1.5 to 4.5 times the global rate. This means that if the current rate of warming continues, we should expect average surface temperatures in the Arctic to increase between 4.2°F (2.3°C) and 12.6°F (7.0°C) over the next hundred years. Most scientists believe that this will result in the Arctic becoming ice free in the summer sometime this century.

The increase also tends to be larger in the winter and spring than in the summer and autumn. And warming causes changes in rain- and snowfall amounts and in the occurrence of droughts, heat waves, storms and myriad other weather events. We'll explore these more fully after we examine the basic physics of **radiation** and the **greenhouse effect**.

THE CONSERVATION OF ENERGY

The conservation of energy is one of the most fundamental laws of physics. What this law means is that energy can neither be created nor destroyed. Rather, it can only be transformed from one form to another. As you might expect, the sun is the ultimate source of virtually all of the energy that we use in our daily lives.

Plants harvest the energy from the sun in a process known as **photosynthesis**. In photosynthesis the sun's energy, carbon dioxide from the atmosphere, and water are combined to form sugar. Oxygen is released back to the atmosphere in the process. During photosynthesis, the sun's energy is transformed and stored as chemical energy. Our body directly taps into this energy when we eat fruits and vegetables.

Burning plants (a process known as combustion) is another way of getting access to stored chemical energy. In combustion, oxygen from the air is consumed and heat, water vapor and carbon dioxide are produced. Similarly, when we burn a **fossil fuel** such as coal, oil or natural gas we are releasing the sun's energy stored many millions of years ago as chemical energy by plants in the ocean and on land. It also took many millions of years for these ancient plants to transform into fossil fuel reserves.

The sun, like everything else, emits energy as **electromagnetic radiation**. Electromagnetic radiation travels at the speed of light. Like ripples on a pond, it has a **wavelength** (the distance between the crests of two waves) and a **frequency** (the number of wave crests that pass a given point in a second). Shorter-wavelength radiation is more energetic than longer-wavelength radiation. Hotter objects emit shorter-wavelength radiation than cooler objects. They also radiate more total energy than cooler objects.

While we may not realize it, electromagnetic radiation is all around us. In the kitchen we often use a microwave to cook our food. When we listen to the radio, we are picking up very long-wavelength and low-frequency radio waves. In the hospital, high-energy X-rays might be used to see if we have broken bones.

When we sit around a campfire, the heat we feel is infrared (long-wavelength radiation) that is being emitted by the burning wood.

Because the sun is extremely hot, with a surface temperature of 9900°F (5500°C), it mostly emits short-wavelength, high-frequency radiation. Our eyes are designed to see this shortwave radiation, otherwise known as visible light. The cooler Earth, on the other hand, emits longer-wavelength, lower-frequency infrared radiation back to space. When the total amount of energy that the Earth receives from the sun is equal to the total amount it emits back to space, we say that the Earth is in **global radiative equilibrium**. This is a fancy way of saying that the Earth's global-average temperature is not changing.

THE GREENHOUSE EFFECT

Most of the atmosphere is made up of nitrogen (78%) and oxygen (21%). Together these gases account for 99% of its volume. The remaining 1% is made up of so-called **trace gases**. These include very small amounts of water vapor, carbon dioxide, ozone, nitrous oxide and methane. Yet without the presence of these gases, there would be no life on Earth. More than 150 years ago, the Irish scientist John Tyndall recognized that these naturally occurring gases were very effective at absorbing the infrared radiation that the Earth emits to space. The absorbed energy is converted to heat, and some of it is reradiated back to the Earth's surface. This is the so-called natural greenhouse effect. These **greenhouse gases** in the atmosphere act like a blanket to keep the Earth's surface temperature warm. If they were not there, the Earth's average surface temperature would be -3.0°F (-19.4°C) instead of the 58°F (14.5°C) observed today. Compare Earth to Mars, which has such a thin atmosphere that its average surface temperature of -58°F (-50°C) is too cold for life.

But too much of a good thing is also bad. Venus has an atmosphere that is 96% carbon dioxide. The surface temperature of Venus is 891°F (477°C), which is too hot for life. Like Baby Bear's porridge in *Goldilocks and the Three Bears*, Earth's temperature is neither too cold nor too hot. It's just right. That is, it has just enough greenhouse gases in the atmosphere to allow life to flourish.

UNDERSTANDING THE GREENHOUSE GASES

Throughout Earth's history the atmospheric concentration of greenhouse gases has varied through natural processes. In fact, a hundred million years ago there was three times as much carbon dioxide in the atmosphere as there is today. The largest natural source of carbon dioxide when averaged over millions of years is volcanic activity. Decaying vegetation is also a natural source. However, if the decay occurs in the absence of oxygen, such as what happens in swamps or wetlands, methane is produced instead of carbon dioxide. Other natural methane sources include termites and deep ocean vents. The most potent greenhouse gas is actually water vapor, which forms when water evaporates. Nitrous oxide is naturally produced in, and released from, soils.

Natural removal mechanisms (known as sinks) also exist for the greenhouse gases. Pure rain is actually slightly acidic because carbon dioxide in the atmosphere can dissolve in water to produce carbonic acid. When rain reacts with certain rocks, small amounts of carbon can be taken up. The carbon is eventually transported to the ocean through river runoff or groundwater. Photosynthesis and the direct absorption of carbon dioxide by the ocean are the two dominant sinks of carbon dioxide. Neither volcanic activity nor the natural removal mechanism of carbonic acid interacting with rocks is relevant to global warming. These processes only affect the atmospheric concentration of carbon dioxide over very long time periods (millions of years). Human activities emit between 100 and 200 times the amount of carbon dioxide released by volcanoes to the atmosphere.

The other main naturally occurring greenhouse gases also have natural removal mechanisms. Methane has a relatively short lifetime of a dozen years in the atmosphere, as it is broken down through chemical reactions. Nitrous oxide is removed from the atmosphere through interaction with sunlight. And water vapor is removed when it rains or snows.

Perhaps the most remarkable and detailed long-term climate records are those obtained from ice cores collected from the Antarctic ice sheet. Teams of scientists have been able to accurately infer levels of atmospheric carbon dioxide, methane and nitrous oxide over the last 800,000 years by directly measuring them in ancient air bubbles trapped in the ice. Prior to **industrialization**, atmospheric carbon concentrations varied between 170 and 300 parts of carbon dioxide per million parts of air by volume (ppmv; see Figure 1.2). Methane varied between 340 and 800 parts of methane per billion parts of air (ppbv), and nitrous oxide between 199 and 303 ppbv. High concentrations of carbon dioxide were typically associated with high concentrations of methane and nitrous oxide, and vice versa.

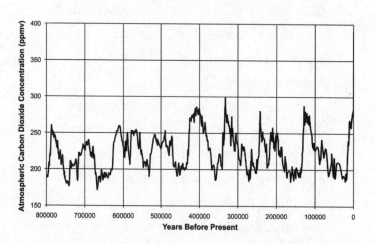

FIGURE 1.2: *This figure shows the variability of carbon dioxide levels over the last 800,000 years. The carbon dioxide concentration is measured in parts of carbon dioxide per million parts of air (ppmv).The present-day level is 390 ppmv. Source: European Project for Ice Coning in Antartica (EPICA).*

To determine atmospheric temperatures over the 800,000-year period shown in Figure 1.2, scientists have used a variety of techniques. As expected, warm climates have high levels of greenhouse gases and cold climates have low levels. At the depth of the last ice age, when most of Canada was buried under thick ice sheets, global temperatures were between 6.3°F (3.5°C) and 9.0°F (5°C) cooler than preindustrial times. During this time the atmospheric concentration of carbon dioxide was about 185 ppmv, methane about 350 ppb and nitrous oxide about 200 ppb.

In essence, these remarkable ice-core observations confirmed the elementary physical principles of the greenhouse effect developed more than 150 years ago. Warm climates can't be maintained unless there is an excess of greenhouse gases to block outgoing longwave radiation; cold climates can't be maintained unless there is a depletion of greenhouse gases. If the amount of these gases is increased, the Earth must warm until a new global radiative equilibrium is reached. The opposite must occur if the amount of these gases is decreased.

So what caused the changes in the atmospheric greenhouse gas concentrations seen in the ice cores? It turns out that the greenhouse gases were naturally released from the oceans, soils and **biosphere** as the temperature warmed. They were naturally taken up by the same as temperatures cooled. That is, the greenhouse gases amplified the warming and cooling that was occurring. And all of this was, in turn, driven by small changes in the properties of the Earth's orbit around the sun that occurred over many thousands of years.

At the time of writing, the atmospheric concentration of methane is 1800 ppbv, more than double anything observed in the last 800,000 years. Similarly, the atmospheric carbon dioxide concentration is 390 ppmv, 30% higher than what occurred over this time period. The atmospheric nitrous oxide concentration is 324 ppbv. Because the oceans respond slowly, it takes several centuries for the Earth to reach a new warmer equilibrium. We know that even if atmospheric carbon dioxide levels were immediately stabilized at 390 ppmv, the Earth would still warm by about 1.1°F (0.6°C) in the years ahead.

SO WHO IS TO BLAME?

Since preindustrial times, human emissions of greenhouse gases have grown very rapidly. This is particularly true for carbon dioxide (see Figure 1.3). The combustion of fossil fuels represents the dominant source of human-produced carbon dioxide. Carbon dioxide is also released when limestone is broken down in the creation of cement (about 4% of the total). Since 1850, humans have emitted about 1932 billion metric tonnes of carbon dioxide to the atmosphere. Almost 70% (1335 billion metric tonnes) of this carbon dioxide is a direct consequence of the combustion of fossil fuels and the production of cement. About 30% has arisen from changes in land use, deforestation in particular.

Humans release methane to the atmosphere through accidental loss during the exploration, drilling, transportation and delivery of natural gas. Methane is produced in landfills when trash decomposes in the absence of oxygen. It can also arise from agricultural activities. Nitrous oxide is largely produced from the heavy use of fertilizers in agricultural activities and, to a lesser extent, from the combustion of fossil fuels. Other important human-produced greenhouse gases are hydrofluorocarbons (HFCs), perfluorocarbons (PFCs) and sulphur hexafluoride. These gases only exist because humans make them. They occur in small concentrations but are very powerful greenhouse gases. Ozone has a short lifetime in the atmosphere (a few weeks). It is not emitted directly but rather arises as a by-product of chemical reactions involving other pollutants such as smog. Human activities produce loads of water vapor in the lower atmosphere. However, the amount that can remain there is tightly controlled by temperature.

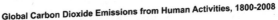

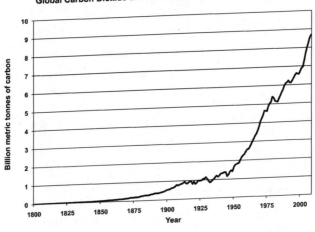

FIGURE 1.3: *Emissions of carbon dioxide as a consequence of fossil fuel combustion and cement production for the period 1800 to 2008. Emissions are expressed in terms of billions of metric tonnes of carbon. In 2008, land-use changes contributed another 1.2 billion metric tonnes of carbon emissions for a total of 9.9 billion metric tonnes that year. To convert carbon emissions to carbon dioxide emissions, multiply the numbers by 3.67.*

Water vapor provides a **positive feedback** to climate change. That is, increasing global temperatures lead to an increase in the amount of water vapor present in the atmosphere. And this in turn leads to further temperature rise. Water vapor does not drive changes in climate. It amplifies the change caused by another process.

Internationally, the United States is the second-largest emitter of carbon dioxide, behind only China (see Table 1.1). Canada ranks in eighth place. However, China's population is more than four times larger than that of the United States. It's also more than forty times larger than Canada's population. This is why China is not ranked very high in terms of emissions per person. Most of the nations with high per-capita emissions are small oil-producing states.

Rank	Nation	Per Capita Emissions	Nation	Total Emissions
1	Qatar	13.46	China	1664589
2	Kuwait	9.35	United States of America	1568806
3	United Arab Emirates	9.00	Russian Federation	426728
4	Bahrain	7.82	India	411914
5	Trinidad & Tobago	6.90	Japan	352748
6	Luxembourg	6.53	Germany	219570
7	Netherland Antilles	6.21	United Kingdom	155051
8	Aruba	6.12	Canada	148549
9	United States of America	5.18	South Korea	129613
10	Australia	4.90	Italy	129313
11	Falkland Islands	4.60	Iran	127357
12	Canada	4.55	Mexico	118950
13	Oman	4.38	South Africa	113086
14	Saudi Arabia	4.38	France	104495
15	Brunei	4.21	Saudi Arabia	104063
16	Faeroe Islands	3.83	Australia	101458
17	Gibraltar	3.65	Brazil	96143
18	Estonia	3.56	Spain	96064
19	Finland	3.45	Indonesia	90950
20	Kazakhstan	3.45	Ukraine	87043

TABLE 1.1: *Left: Top 20 countries in 2006 ranked by annual emissions of carbon in metric tonnes per person (one metric tonne is one thousand kilograms). Right: Top 20 countries in 2006 ranked by total annual carbon emissions in thousands of metric tonnes. Canada: light gray shading; United States: dark gray shading. To convert carbon emissions to carbon dioxide emissions multiply the numbers by 3.67. Source: Carbon Dioxide Information Analysis Center (CDIAC).*

So where do our emissions come from? A little more than a quarter of all greenhouse gases in Canada and the United States come from the transportation sector (see Figure 1.4). This includes the automobiles that we drive to and from work and school. Thirty-five percent of US emissions come from the generation of heat and electricity. And 80% of that is from coal-fired electricity production. In Canada, emissions from the fossil fuel industry account for 22% of the total. This is also the sector in Canada with the most rapid growth in emissions. The agriculture and waste sectors together contribute only 8% of total emissions in the United States and 11% in Canada.

Knowing the largest sources of greenhouse gases is important as it lets us know where the biggest reductions have to occur. As an example, the United States would make substantial emissions reductions by replacing coal-fired electricity production with renewable or nuclear energy. Obviously this assumes that as a society we want to deal with global warming.

The global emissions of carbon dioxide are showing no signs of decreasing. Presently atmospheric carbon dioxide levels are increasing by about 2 ppm per year. If this

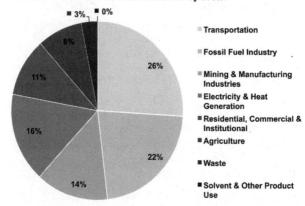

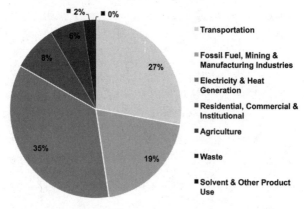

FIGURE 1.4: *Percentage of total emissions by sector in 2008 in Canada (above) and the United States (below). The Fossil Fuel, Mining and Manufacturing Sector emissions have been broken down into two categories (Fossil Fuel, and Mining and Manufacturing) in the case of Canada. Source: Environment Canada and the United States Environmental Protection Agency (EPA).*

rate of increase doesn't change over the next ninety years, the carbon dioxide concentration will reach about 570 ppm by the end of this century. In actual fact, there is a wealth of evidence to suggest that both the ocean- and land-based carbon sinks become less effective as the climate warms. This is why some estimates have the atmospheric carbon dioxide level reaching as high as 1000 ppm this century if we fail to take action.

Atmospheric carbon dioxide levels have been much higher than they are today throughout much of Earth's history. The difference between now and the past is the rate at which carbon dioxide levels are changing. The coal, oil and natural gas reserves found on Earth today were formed over tens of millions of years, storing carbon dioxide in the process. Now we are turning back the clock and returning that carbon dioxide to the atmosphere in a few decades.

There is no evidence to suggest that humans are kicking off a runaway greenhouse effect that will turn Earth into a planet like Venus. We know that life has flourished on Earth in the past with much higher atmospheric carbon dioxide levels than today. But that life was quite different from what exists now. There is overwhelming evidence to suggest that the rapid rate of warming will lead to mass ecological extinctions as various species fail to adapt.

CLIMATE MODELS

You might be wondering how scientists are able to tell that most of the twentieth-century warming is a consequence of human activities and not just a result of natural climate change. There are two main drivers of natural climate change: volcanoes and the sun. When a volcano erupts, it emits tiny solid and liquid particles into the atmosphere. These particles affect the amount of sunlight that reaches the Earth's surface by reflecting some of it back to space. They have a net cooling effect. But they don't last very long in the atmosphere. They are washed out when it rains or snows. It's only when the particles are injected into the upper atmosphere by very powerful volcanoes that they have a longer-lasting effect. And even then the cooling is only temporary, lasting a year or two at most. The most recent example occurred when Mount Pinatubo in the Philippines erupted in June 1991.

Estimates of variations in the strength of the sun's output have existed for some time. Since 1979, direct measurements have been available from orbiting satellites. While there is no doubt that changes in the sun's intensity affect climate, in the last few decades the sun has been cooling at precisely the time when the Earth's warming has been accelerating.

Our scientific understanding of the physics of the environment has advanced systematically and profoundly since the seventeenth-century work of Isaac Newton. The work of thousands of scientists has been published in scientific journals over the years. Common to these studies is the use of the scientific method. Similar advances in chemistry, biology and the other branches of the natural sciences have also occurred. Climate models have been developed as a way of integrating this knowledge. They simultaneously solve complicated equations governing the movement of, and physical and chemical interactions between, the atmosphere, ocean, sea ice and land-surface ice, including glaciers and ice sheets. These equations are simply too complicated to be solved with a pencil and paper. They require a computer.

We can imagine how a climate model works by thinking of a scaled replica of the Earth, its ocean and overlying atmosphere made from Lego bricks. Each brick is in contact with a number of neighboring bricks.

Each of these bricks is, in turn, in contact with other bricks. We have several layers of bricks representing the land surface, its subsurface properties and any glaciers or ice sheets that may be present. Sea ice requires another few layers of bricks sandwiched between the atmosphere and ocean. The bricks themselves don't move. Rather, they exchange heat, moisture, carbon and momentum with neighboring bricks. And these quantities must be conserved. They can't be created or destroyed. There are millions of equations representing the interactions of all these Lego bricks that must be solved.

Before a climate model is used to try and understand what is in store for the future, it has to be extensively tested. The model must be able to reproduce present and past observations of the climate system. When initiated with preindustrial conditions, it has to be able to reproduce the observed twentieth-century climate. This requires checking that when observed changes in volcanic emissions, solar intensity and human emissions are prescribed, the model simulates the evolution of the twentieth-century climate.

What climate models tell us is that there is no way of explaining the late-twentieth-century warming through any known mechanism of natural variability. These same models tell us that when we include the effects of human activity, we can explain this warming. This is called climate change detection and attribution.

These same models are used to project as to what might happen to the climate as a consequence of human activities. There are a number of groups internationally that have developed climate models. The two main groups in the United States are at the National Center for Atmospheric Research in Boulder, Colorado, and the NOAA Geophysical Fluid Dynamics Laboratory in Princeton, New Jersey. In Canada, the main group is Environment Canada's Canadian Centre for

Climate Modelling and Analysis, located on the campus of the University of Victoria.

THINGS ARE HEATING UP

It's important to remember that the projections from climate models are not weather predictions. Rather, they tell us about changes in the background climate state on top of which daily weather occurs. They also inform us of changes in the likelihood of any particular weather event.

Climate models tell us that we should expect somewhere between 2.0°F (1.1°C) and 11.5°F (6.4°C) of global warming by the end of the twenty-first century. Of course, the overall warming is sensitive to our future emissions of greenhouse gases. If emissions continue to grow through our use of fossil fuels (especially coal), then we can expect greater warming than if we maintain or reduce our emissions. This may not seem like a lot.

But remember that the world was only about 6.3°F (3.5°C) to 9.0°F (5.0°C) colder than today in the depths of the last ice age. Global warming of more than about 3.6°F (2.0°C) has not been seen in at least the last 2.6 million years, and the projected rate of warming over the next several centuries is unparalleled in the last 50 million years.

The warming will be largest at high latitudes, where strong positive feedbacks exist. As snow and sea ice melt, the surface becomes darker. Dark surfaces absorb more incoming solar radiation than white surfaces. This absorbed radiation is converted to heat. In addition, as sea ice melts, an insulating buffer between the ocean and atmosphere disappears. This means that heat from the ocean can more readily warm the lower atmosphere.

Also, land warms up faster than ocean. In fact, it takes about five times as much energy to warm one gram of water as one gram of dry soil. Many of us have witnessed this firsthand when we go to the beach in the summer. The sand is hot and the water is cool. Locations near the coast warm more slowly than inland areas. This follows since the air that blows over coastal regions is strongly influenced by water upwind. Because there is more land in the Northern Hemisphere, it warms more than the Southern Hemisphere.

As a consequence of the overall warming, we should expect warm temperature records to be broken more often than cold records. Recent research for the United States has shown this to be the case. Record warm temperatures are being broken at about twice the rate of record cold temperatures. By the mid-century, it is expected that this will increase to about twenty to one. If emissions remain high for the remainder of this century, we can expect a further increase to as much as one hundred to one. What this means is that the odds of breaking a warm record any given day would be one hundred times the odds of breaking a cold record that day. We will explore the consequences of these changes in temperature extremes in Part Two.

IT TAKES TIME TO WARM

As I mentioned earlier, the response of the climate system to an increase in greenhouse gases is not immediate. It takes time for the oceans to warm up as they absorb carbon dioxide. The analogy to a pot of water on a stove is illustrative. When you turn on the stove's element, it takes time for the water to warm. Similarly, when you turn off the element, it takes time for the water to cool. It's the same with the climate system. If we keep atmospheric greenhouse gases at present levels, we still have about as much additional warming in store as has already occurred since preindustrial times. It takes time for the Earth to respond to present-day greenhouse gas levels. Climate model simulations tell us that if we burn existing fossil fuel reserves, about two-thirds of the resulting increase in temperature will last more than 10,000 years.

SEA LEVEL IS RISING

As the oceans warm, the water expands. This causes sea level to rise. Another major component of sea-level rise comes from the melting of glaciers and ice sheets on land. (Incidentally, the melting of floating ice does not cause changes in sea level since the mass of the floating ice is already displaced). Observations suggest that the global sea level is presently rising by about 0.13 inches (3.3 millimeters) per year. About 40% of recent sea-level rise has come from the expansion of warming waters and about 60% from the melting of ice on land. Recent analysis has suggested that we should expect somewhere between 2.0 feet (0.6 meters) and 5.2 feet (1.6 meters) of sea-level rise by the year 2100. But as there are for temperature, there are substantial regional variations. The northeast coast of the United States is expected to receive greater sea-level rise than the global average due to changing ocean circulation in the North Atlantic.

Not surprisingly, most of the world's glaciers and ice caps are shrinking. In addition, both the Greenland and Antarctic ice sheets are melting. Recent research suggests that with warming of more than 3.6°F (2.0°C) since preindustrial times, there is a very high probability that the Greenland ice sheet will become committed to disintegration. This would give a sea-level rise of about 22 feet (6.6 meters) over the next several centuries. Less is known about the stability of the West Antarctic ice sheet. However, some have argued that 3.6°F (2.0°C) would push it perilously close to, if not past, the point of no return. The most vulnerable part of this ice sheet would contribute a further 11 feet (3.3 meters) of sea-level rise. If the entire West Antarctic ice sheet were to collapse, it would lead to about 16 feet (5.0 meters) of sea-level rise.

PRECIPITATION PATTERNS ARE CHANGING

Climate models predict that changes in the amount and type of precipitation will occur. Many of these changes have already begun around the world. As the world warms, regions that are typically wet today will get wetter. Those that are already dry today will get drier. When it rains or snows, there is a greater likelihood of the rain or snow falling in heavier amounts.

Climate models are able to quantify this change. For example, depending on our future emissions, these models estimate that by the end of this century the once-in-twenty-year rainstorm will occur once every seven to thirteen years. As such, there will be an increased risk of flooding. Of course, the likelihood of the precipitation falling as rain instead of snow increases as the century progresses. So while we will never be able to say that global warming caused the 2010 record-breaking flooding in Pakistan, we can say that the likelihood of more and even bigger flooding events will increase as the century progresses.

In Canada and the northern United States, overall precipitation will increase, but it will come in fewer, more extreme events, interspersed with longer periods of little or no precipitation. The precipitation will be skewed to the winter. And summer drought will become more common. While there will be lots of water around, it will come at times when it's not needed and not at times when it is needed. Global warming will create an issue of water storage, not water availability, for these regions.

At the same time, the southern states will become drier. This will create a greater pressure on already depleting aquifers.

Global warming will create a water-shortage issue for these regions. There will likely be pressure to transport water from northern regions of North America to the more southern regions. Desalination plants may need to be built to extract freshwater from seawater.

As the Earth warms, the difference between polar and subtropical surface temperatures decreases, especially in the winter. This occurs because warming is amplified at high latitudes. Since midlatitude storm formation is sensitive to the strength of the north-south temperature difference, we can expect fewer storms. But at the same time, we can expect an increase in the likelihood of stronger storms. That is, while the total number of storms at middle latitudes goes down, the number of stronger storms goes up. And these strong storms are associated with stronger winds. The overall number of tropical cyclones (called hurricanes in the Atlantic) is expected to decrease slightly. But again, there is an expectation that the number of stronger Category 4 or 5 storms will increase as the century progresses.

ECOSYSTEMS IN TROUBLE

The single biggest natural sink for the excess carbon dioxide we put into the atmosphere is the ocean. As the ocean slowly takes up this carbon, the water becomes more acidic. Corals, certain types of plankton, mollusks such as mussels or oysters, and crustaceans such as crabs or lobsters all have shells made of calcium carbonate. The formation of these shells is very sensitive to the amount of carbon dioxide dissolved in the ocean. Many species will find that it will be difficult for them to grow shells. At high-enough levels of carbon dioxide, the shells of some living species will actually start to dissolve.

We can do a simple experiment at home to watch this in action. Drop a piece of chalk into carbonated water. The chalk, which is made of calcium carbonate, starts to dissolve. That being said, some recent research has suggested that there are a few organisms, like the American lobster or blue crab, that grow better in seawater with more carbon dioxide. However, even these creatures may suffer if the ecosystems upon which they rely and feed respond negatively to increasing acidity levels.

Once more, it is the rate of change in ocean acidity that poses the biggest problem for marine organisms. The current rate is unparalleled in the last 65 million years. It is ten times faster than occurred 55 million years ago during a major deep-ocean extinction event. In addition, if we burn existing fossil fuel reserves, by the end of the twenty-second century, ocean acidity will reach levels that have not been seen for at least 300 million years. It takes thousands of years for atmospheric carbon dioxide levels to drop, and many thousands more for the ocean to become rebalanced. Increasing ocean acidity, with its potential effect on resident ecosystems, is an extremely serious indirect consequence of our increasing atmospheric carbon dioxide emissions.

Ecosystems on land are also vulnerable to the effects of global warming. Recent estimates point out that further warming of about 1.6°F (0.9°C) from today will cause between 9% and 31% of the world's species to become committed to extinction. If global temperatures rise 5.4°F (3°C) above today's values, more than half of the world's nature reserves will no longer be able to fulfill their conservation mandates. Finally, with future global warming of about 6°F (3.3°C), the best estimate is that between 40% and 70% of the world's species will become extinct. Climate models suggest that our business-as-usual approach to fossil fuel consumption is currently on track to take us to beyond 6°F global warming.

None of this means that life will cease to exist on Earth. Life has flourished over most of its history and especially over the last half billion years. Earth has also previously undergone a number of life-crippling extinction events associated with asteroid impacts or decades of massive and sustained volcanic activity. But more often than not, the life that reemerges after an extinction event is quite different from the life that was present before.

PART TWO

Why Should I Care?

As people go about their daily lives, it's unlikely they spend much time wondering how global warming is affecting them now or how it will affect them in the years ahead. There are far more urgent matters to attend to. Children need to be taken to school. The bills have to be paid. There are pressures at work and deadlines to meet. For many, global warming is an abstract problem. It's difficult to imagine how a few degrees of warming over the next hundred years will affect us today or, for that matter, anytime in the near future.

The results from climate models compli-cate matters. They tell us that the amount of warming over the next century is very sensi-tive to our future emissions of greenhouse gases. But they also tell us that this is not the case for warming over the next several decades. The projected climate change for the next twenty to thirty years is very similar whether we continue with growing emissions or we start to stabilize and slowly reduce emis-sions. This means that global warming is an issue that spans generations.

So in some sense, our decision to deal with global warming or not boils down to one ques-tion: Do we have any responsibility for the well-being of future generations? Unfortunately, science cannot provide an answer to this ques-tion. It must be answered by society as a whole. Are we willing to take the steps required today to ensure that future generations are able to enjoy the same economic stability and ecolog-ical diversity that we currently enjoy?

It's no wonder that our political leaders are having such a difficult time introducing the policies needed to ensure a reduction in greenhouse gases. Politicians are typically elected for short terms in office. Every four years or so there is a new election. Let's suppose that there is a health-care crisis in a particular city. A politician may get elected on the grounds that he or she will deal with this crisis. A hospital might get built. During the next election campaign, the politician can point to the hospital and say to his or her constituents: "Look. I listened to you. We built a hospital to deal with your local health-care problem." That politician may get reelected. Now let's suppose you are a politician who introduces a regulation limiting greenhouse gases. Or you might add a tax or levy to greenhouse gas emissions. The effects of this policy would not be realized during your political career. In fact, they may not be realized in your entire lifetime. They would start to have an effect in the lifetime of the next generation. That's hardly something you can point to in the next election campaign. There is no immediate benefit.

THE TRAGEDY OF THE COMMONS

Back in 1833, the British political economist William Forster Lloyd published *Two Lectures on the Checks to Population*. He discussed an example where cattle grazed on either private or public lands. Lloyd noted that when a farmer sells a cow, he gets all the profits no matter where his cow grazed. If the farmer owns the land on which the cow fed, he would pay the entire cost for any overgrazing that led to smaller, weaker cattle in his herd. The farmer wouldn't keep adding cows to his pasture, because at some point he would realize that he has nothing to gain. Now let's suppose that the farmer grazes his cattle on public land shared with many other farmers. Each extra cow that feeds off the land will wholly benefit its owner upon sale. But the costs of overgrazing would be borne by all farmers. In this case, it's to the farmer's advantage to add an extra cow to the pasture since he only pays a fraction of the cost of doing so. But what if all farmers are thinking the same thing? It's in all of their individual interests to add more cows. The finite size of the pasture means that cows can't be added forever. The grass will run out and the cows will starve. An eventual collapse must occur. This was termed "the tragedy of the commons" by ecologist Garrett Hardin in 1968.

The tragedy of the commons has a direct connection to global warming. There is no incentive for any country, state, province, city or individual to reduce greenhouse gas emissions. The entire cost of reduction would be borne by the individual, city, province, state or country. The cost of inaction would be distributed globally. The status quo prevails until collapse ensues.

There is no doubt that limiting global warming will be difficult. The tragedy of the commons tells us that a solution to the problem will require people to think about the broader consequences of their individual actions and decisions. It tells us that technology alone will not solve things. This is because we would only adopt new technologies if we would immediately benefit from them. Why pay more for renewable electricity if it can be produced so cheaply from the burning of coal? The tragedy of the commons tells us that widespread technological innovation must be coupled with behavioral shifts. And these will likely only occur through education.

For the remainder of Part Two, I will describe some of the projected impacts of global warming this century. My goal is to arm you with enough information to decide whether or not you think we should take steps to reduce global warming. In Part Three, I will assume that you think we should, and I will suggest some ways to move forward.

HOW CLIMATE AFFECTED OUR ANCESTORS

Humans have been on Earth for almost 200,000 years. Our early ancestors relied upon hunting and gathering to feed themselves. At the end of the last ice age, about 11,000 years ago, human food production underwent a revolution. It was at this time that agriculture and the domestication of animals evolved. Ice cores and other records tell us that over the last 11,000 years the climate has been relatively stable. Global average temperatures have varied by only about, plus or minus, 1°F (0.5°C). That's not to say climate hasn't changed over this time. It has. But the changes were small enough to allow human civilization to flourish through its dependence on agriculture and livestock. This was not the case prior to the end of the last ice age.

Changes in climate have caused societies to rise and fall. For example, the Vikings settled the southern tip of Greenland in AD 984 during a time known as the Medieval Warm Period. Global temperatures during this period were about 1°F (0.5°C) colder than the average over the first decade of the twenty-first century. Almost a century earlier, the Mayan civilization in southern Mexico collapsed following a sequence of extended droughts. The Viking settlement also died off in the 1420s during a time known as the Little Ice Age. Of course, today we would have less difficulty than the Mayans or Vikings in adapting to the changes in climate that they faced. We have access to incredible advances in technology. Nevertheless, both the magnitude and the rate of change of climate over this century will be far greater than that faced by any of our ancestors over the last 11,000 years.

ADAPTING TO CLIMATE CHANGE

We know that we will have to adapt to the global warming that will occur over the next several decades. This is termed *climate change adaptation*. But what we need to adapt to after that depends on the steps we take to reduce greenhouse gas emissions. Reducing emissions is called *climate change mitigation*.

In the years ahead we will have to deal with an increase in extreme precipitation. In winter this translates to an increased likelihood of very large snowfalls. City planners will need to design storm-drain systems carefully to handle bigger downpours. New approaches to river floodplain management might be required. Agricultural practices will need to be revisited in light of changing precipitation patterns.

Adaptation will also be required to maintain human comfort levels. As the Earth warms, summer heat waves will become more common. This means rising demand for air-conditioning. More electricity will be needed in the summer. Conversely, warming winters will lead to a reduction in home heating costs.

Sea level will slowly put pressures on low-lying coastal communities. Coastal erosion and inland flooding will become increasingly problematic for these regions. Some regions will be forced to build new or bigger dikes and levees. In the Arctic, melting permafrost will pose problems for buildings and roads that will slump under the softening ground.

It's hard to do long-term planning for climate change adaptation unless you know in advance what measures will be implemented to mitigate climate change. We can illustrate this using two examples.

First let's imagine that a new hydroelectric dam is going to be built. We know that it will be very expensive to build. It is also meant to last for many decades. For example, the Grand Coulee dam on the Columbia River is already more than seventy-five years old. We could build the new dam assuming that rain and snow amounts will always be the same as today. That will almost certainly be a bad decision. Depending on where the dam is built, there could be more or less rain or snow in the future. We know that as the century progresses the likelihood of precipitation falling as rain instead of snow will increase. We also know that the spring melt season will come earlier in the year. This means that both the amount of water and the time when it reaches the dam will change. So if you are going to spend six billion dollars constructing a new dam (the current estimate for the cost of the Site C dam on the Peace River in British Columbia), you had better make sure there will be enough water to power it at the time of the year when the electricity is needed.

Now let's examine decision making at a forestry company. After a forest is harvested, it needs to be replanted. Decisions must be made as to the type of tree that will grow both now and into the future. Will there be enough water for the trees to grow? What will the temperatures be like? Will the changing climate support more pest or fire outbreaks? If fast-growing poplar trees are to be planted, they will be harvested in about thirty years. However, slow-growing Douglas fir, white spruce or oak trees wouldn't be harvested for over one hundred years. The climate will change considerably over this time, depending on what, if any, mitigation is done.

Some economists have argued against taking steps to reduce greenhouse gas emissions. They suggest that we would be wiser to use our resources to grow our economies. We would then become wealthier. This would better position us down the road to pay for any adaptation costs or damages associated with global warming. Economists term this *discounting* future costs.

Let's suppose you have an old roof on your house that you are convinced will last five more years. You have wisely saved $10,000 in the bank to pay for the roof's replacement. But you wonder whether you should spend the $10,000 now or later. If you leave the money in the bank, it might earn 5% interest compounded each year. In five years it will grow to $12,763. Inflation has been low lately. Projections are that it will remain at about 2% for the foreseeable future. If it cost you $10,000 now to replace the roof, you would expect it to cost you $11,041 to replace it in five years. Economically, you would be better off waiting five years to replace the roof since you would be $1722 ahead. From this example you might correctly conclude that a dollar today is worth more than a dollar tomorrow. Of course, you would have made the wrong decision if the roof leaks in year three and causes $20,000 in uninsured damages inside your house.

The typical economic discounting of future costs does not work unless future changes occur in a fairly orderly fashion. Also, some of the biggest global warming impacts will occur in some of the poorest parts of the world. This of course leads to the question of who is discounting whose costs? The developed world discounts its cost of action against the costs of inaction somewhere else in the world. Issues of fairness get raised. Discounting costs decades ahead also ignores the potential rights of future generations. Do we have the moral right to burden our children and our children's children with these costs? The real truth is that economic arguments for inaction now versus action later are far more complicated than many are led to believe.

GLOBAL WARMING AND OUR HEALTH

Global warming is projected to have both positive and negative effects on human health.

Summer heat-related mortalities are expected to increase. An extreme example of this occurred in Europe during August 2003. Thirty-five thousand people died from heat-related causes. At the same time, the number of deaths from exposure to cold in the winter is expected to decrease.

When and where infectious disease outbreaks occur will also be affected by global warming. The negative effects will be largest in some of the poorest, less-developed parts of the world. These same regions already struggle with a higher incidence of infectious diseases than North America. Some infectious diseases will expand into new areas. Warming temperatures in North America could lead to a reduction in the occurrence of respiratory diseases such as the flu and pneumonia.

West Nile virus was first discovered in Uganda in 1937. This tropical disease entered the United States in 1999. By 2010 it had firmly taken hold throughout the continental United States and southern Canada. Warming temperatures and the earlier arrival of spring have been implicated in the cause of the outbreak. Another mosquito-borne disease that will likely move northward is dengue fever. This nasty and sometimes fatal illness showed up in Florida in 2010. Outbreaks occasionally occur on the Texas border with Mexico.

Lyme disease, carried by ticks, is also becoming more common as the climate warms. Cold winter temperatures are needed to kill off the ticks. Lyme disease should become well established in southern and eastern Canada in the years ahead. It is already firmly entrenched in British Columbia.

Many water- and food-borne diseases are also susceptible to changing climate. Increasing extreme precipitation and subsequent flooding has led to outbreaks of cholera in parts of Africa. Closer to home, *E. coli* outbreaks have been linked to extreme precipitation events. For example, 5.3 inches (134 millimeters) of rain fell at Walkerton, Ontario, over a five-day period from May 8 to May 12, 2000. Of this, 2.8 inches (70 millimeters) fell on May 12. A total of 2321 people became ill, and seven eventually died from an *E. coli* outbreak caused by the runoff from farms infiltrating the water supply. Among other diseases, *Salmonella* is a type of bacteria that can lead to food poisoning. It is known to reproduce more rapidly in warmer conditions.

One of the consequences of global warming for Canada and the northern United States is that precipitation will come in fewer, more extreme events, interspersed with longer periods of little or no precipitation. Extended dry periods allow tiny particles, including pollen and pollution, to remain in the air longer. As a result, the incidence of asthma, allergies and hay fever will likely increase. In 1999, a tropical fungus, *Cryptococcus gattii,* was found to have established itself in some soils and trees on Vancouver Island. Previously it only existed in places like Australia and Papua New Guinea. Since 1999, the fungus has spread, and a new strain has emerged in Oregon and Washington state. Humans are affected when its spores are inhaled. The fungus can then infect the lungs. Infectious-disease specialists have attributed the emergence of *Cryptococcus gattii* to warmer and drier summers that have characterized the Pacific Northwest in recent years.

ECOSYSTEMS IN MORE TROUBLE

One of the most profound consequences of global warming will be the toll it takes on many of nature's species, both on land and in the ocean. The magnitude and the rate of climate change over this century will be far greater than most of these species have had to adapt to in the past. They have two options. They either learn to adapt or become extinct.

In North America many plant species are already adapting to the earlier arrival of spring by flowering early. Some birds are nesting earlier; some animals are breeding earlier. In response to warming temperatures, many species are shifting northward and to higher elevations.

Warming winter temperatures have led to an increase in pest outbreaks in North America's northern forests. Normally, cold winter temperatures are effective at killing off insect larvae. But in British Columbia a series of warm winters led to a massive outbreak of the mountain pine beetle. The result is that somewhere between 50% and 80% of British Columbia's mature lodgepole pine forests are dead or dying. Similarly, in Alaska the spruce budworm has halved the time it takes to complete a life cycle in recent years. This has led to an outbreak. Once more, warming temperatures have been implicated as the cause.

While warming winter temperatures have led to increased pest outbreaks, warming spring and summer temperatures have led to an increase in the number and extent of forest fires. In the western United States there has been a 400% increase in the number of forest fires and a 650% increase in the area burned compared to the period 1970 to 1986. Similar increases are seen in western Canada.

Left unchecked, global warming and increasing ocean acidity have the potential to create widespread species extinction. Society as a whole needs to address the question of how much we value these natural ecosystems. It's not possible simply to assign a dollar value to them. There are ethical and moral judgments that must be made. Do we care if polar bears, frogs, butterflies or corals go extinct? What do we see as our own role as a species on Earth? Should we consider ourselves caretakers or stewards of the environment?

Are Earth's natural resources here exclusively for our use and exploitation? Can we, or should we, live sustainably?

TIPPING POINTS IN THE CLIMATE SYSTEM

In 2000 Malcolm Gladwell published *The Tipping Point.* In it he described the existence of critical moments in time after which a cascade of change occurred. One of the many examples he used to illustrate the concept was the cell phone. The first cell phones were big and expensive, and the coverage was poor. During the early 1990s the price started to come down. The phones got smaller. More and more towers were installed. Coverage got better. In 1998, a **tipping point** was reached. The low price, small size, convenience and widespread coverage meant everyone started buying cell phones. The introduction of the flat-screen TV into the market is another example.

The climate system also contains tipping points. With further global warming of between 2.2°F (1.2°C) and 7.0°F (3.9°C), the Greenland ice sheet passes such a threshold. Beyond this tipping point we become committed to an eventual 22 feet (6.6 meters) of sea-level rise as the ice inevitably melts away. Fortunately the rise doesn't happen overnight. It takes many centuries to occur. A tipping point also exists for the West Antarctic ice sheet, although its exact value is more uncertain. Some estimates have it occurring with further warming as low as 2.3°F (1.3°C). An additional 16 feet (5 meters) of sea-level rise would occur as a result of the collapse of the West Antarctic ice sheet. Again, this would take many centuries to be realized.

Yet another example of a tipping point concerns the **dieback** of the Amazon rain-forest. Estimates suggest that global warming of between 5.4°F (3.0°C) and 7.2°F (4.0°C) will

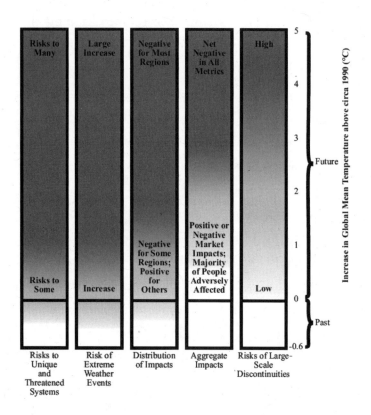

FIGURE 2.1: *Potential reasons for concern about global warming. Global warming impacts are plotted against the future increase in global average temperature in °C from 1990. A Fahrenheit scale can be approximated by multiplying the temperatures by two. Shown is a 0.6°C (1.1°F) increase in global average temperature between preindustrial times and 1990. Darker shading means greater risk. "Risks of Large-Scale Discontinuities" means the risk of passing a tipping point. Reprinted with permission of JB Smith and colleagues (2009) Proceedings of the National Academy of Sciences, 106 (11), 4133–4137.*

lead to a dramatic dieback of the rainforest. Other potential tipping points involve changes in the Indian and West African monsoon season, ocean circulation and the boreal forest. In all cases, passing a tipping point leads to irreversible change.

A summary chart of potential reasons for concern about global warming is shown in Figure 2.1. As the Earth continues to warm, we increase the risk of negative impacts. While initially some northern countries might actually benefit from global warming, the impacts eventually become negative for most nations. Extreme weather events such as floods, droughts and heat waves increase as the Earth warms. Warming leads to stress on unique and threatened systems, such as coral reefs and glaciers; many species face extinction. The total costs of global warming increase with temperature.

Given the scale of some of these potential consequences, you might agree with the application of the **precautionary principle**. Let's suppose that you are unconvinced that there is a scientific consensus as to the causes and consequences of global warming. You would still have to agree that there is a *potential* risk of harm to the general public. Many scientists have expressed this risk as individuals, through their published scientific findings and through formal national and international assessments. When such a potential risk exists, the precautionary principle would imply that measures should be taken to reduce greenhouse gas emissions. This is because the precautionary principle requires us to prove our actions will *not* cause a problem instead of proving that a problem will exist.

GLOBAL SECURITY

Our ability to adapt to global warming depends not only on the rate of change and magnitude of the warming but also on our wealth.

Some of the biggest projected impacts will occur in some of the poorest parts of the world. Yet these same regions have contributed little to the past emissions of greenhouse gases. It's easy to imagine that large parts of the world will develop resentment toward western nations.

The fact that western economies depend so much on foreign oil is troublesome. Many oil-producing countries are politically unstable. Growing demand is also being placed on dwindling reserves. This demand will only increase as India and China continue to develop economically.

The impacts of global warming will lead to the emergence of a new brand of environmental refugee. While so many of us take the availability of fresh water for granted, there are large parts of the world where it is in short supply. As the twenty-first century progresses, water will become scarcer in the subtropics. Millions of people will likely be forced to move in search of food and water. This will spark regional tensions. Wealthy coastal nations might respond by building factories to extract freshwater from seawater. But this will be impossible for landlocked countries and difficult for poorer nations. Low-lying coastal areas and some small island nations will also be under pressure from rising sea levels.

Environmental refugees are already leaving the Lake Chad region of Africa. The lake straddles four countries: Niger, Chad, Cameroon and Nigeria. It has been reduced in size by 90% over the last four decades and will almost certainly dry up in the next few decades. Already there are tensions between local herders, fishers and communities as they compete for dwindling resources. These tensions are likely to grow.

There have also been environmental refugees from a few small, low-lying islands in recent years. In 2005 about a thousand residents were evacuated from the Carteret Atoll in Papua New Guinea. Rising seas were flooding their land. Closer to home, the town of Kivalina in Alaska is facing an uncertain future. The island on which the town is built is eroding away as the underlying permafrost melts. Residents of Kivalina actually filed a lawsuit against eighteen oil companies. They claimed that these energy companies caused global warming that led to the town sustaining damages. A similar court case was started by a number of residents of the Mississippi Gulf Coast who were affected by Hurricane Katrina in 2005. They argue that greenhouse gas emissions contributed to the hurricane being so strong.

The island nations of Tuvalu, the Maldives, Kiribati and the Marshall Islands are also in serious trouble. Most of their area is at, or only a meter or two above, sea level. It is unlikely that these countries will be habitable toward the end of this century or in the twenty-second century. The government of Tuvalu has already entered into negotiations with New Zealand about the possibility of relocating its twelve thousand residents should the need arise.

One of the greatest potential consequences of global warming is therefore its effect on global security. Dwindling water resources in parts of the world will spark regional tensions. Some areas will become uninhabitable. The best estimate of the United Nations International Organization for Migration is that there will be 200 million environmental refugees from global warming by 2050. Reducing our consumption of fossil fuels has many co-benefits for global security. We reduce our dependence on foreign oil from politically unstable parts of the world, and we also reduce the number of environmental refugees.

SO MANY BENEFITS TO CHANGE

The solution to the global warming problem involves two key elements. These are technological innovation and behavioral change.

Together they will allow us to transform our energy systems away from burning fossil fuels toward the use of renewable resources. Doing so has many co-benefits beyond simply national and global security. These include new business opportunities and jobs as society moves to a sustainable low-carbon future. At the same time, parallel measures taken to conserve energy also save money.

Many air pollutants, including smog, are by-products of burning fossil fuels. These are notorious for causing respiratory problems, especially for people with asthma. Burning fossil fuels is also noisy. Imagine driving around in electric cars. There would be no tailpipe emissions. Noise pollution would be greatly reduced. You could even fill up your car at home. Acid rain is also a consequence of fossil fuel burning. It too would be eliminated.

Eliminating deforestation helps preserve the diversity of nature. Perhaps the cure for cancer lies within the flowers of a yet-to-be-found tropical plant. At the same time, soil erosion and unwanted runoff are reduced. Forests lock up carbon and help regulate Earth's temperature.

The cartoon shown in Figure 2.2 sums it up. Even if you don't believe we have any responsibility for the well-being of future generations, you might want to reduce greenhouse gas emissions because of the many other benefits of doing so.

FIGURE 2.2: *Joel Pett editorial cartoon originally published on December 7, 2009. Reprinted with permission from the Cartoonist Group.*

PART THREE

What Can I Do About It?

We're now at a point where each of us has to ask ourselves three questions. The first question is this: Do we have any responsibility for the well-being of future generations? I've tried to provide enough background information in Parts One and Two to illustrate what these future generations might have to face. If the answer to this question is yes, then the second question is: What can I do about it? Since a certain amount of warming is inevitable, the third question needs to be addressed no matter how you answer question one. What can we do to adapt to global warming?

Many communities, states and provinces have either developed or are in the process of developing climate change adaptation plans. These are usually easy to find on the Internet. Typically these jurisdictions start by assessing the threat of climate change to their physical, social and natural environment. For example, a particular local ecosystem might be unable to tolerate drought. An adaptation strategy would be finding a way to sustain this ecosystem in a changing climate. A seaside tourist resort may be concerned about rising sea levels and coastal erosion. A logging community might ponder its future if increased pest outbreaks or forest fires were to decimate local forests. Climate adaptation plans are also extremely important in the design of new infrastructure that is expected to last for many decades.

Solutions to the global warming problem generally fall into two categories: technological and behavioral. The technological challenge is enormous. We need to move our energy systems away from fossil fuels. Of course there are many benefits to doing this anyway, including the fact that oil is a depleting resource.

The technological barriers may be large, but, in truth, most of the solutions are readily available. They are just costly. As we shall see next, the single most important solution to global warming involves putting a price on greenhouse gas emissions. Pricing emissions levels the playing field. New energy technologies are then able to compete with traditional fossil fuels. As they become commonplace, their price goes down. Pretty soon almost everyone can afford them, just as they can now afford cell phones and flat-screen TVs.

Behavioral barriers are also present. Our current patterns of consumption are unsustainable. This doesn't mean we should all stop buying cars, flat-screen TVs or cell phones. Rather, it requires us to think more deeply about how these are produced and how they will be disposed of when they are no longer working. As a society we will need to evolve from *Generation Me* to *Generation Us*. In recent decades North Americans have not had to live in a world where duty and greater good are placed before personal entitlement and individual needs. Instead of fixating on how the actions of others affect us as individuals, we will be compelled to focus on how our actions as individuals affect others. And we must move away from a culture of fear and denial to one of excitement and empowerment. It's a tall order. But it is within our grasp.

PUTTING A PRICE ON GREENHOUSE GAS EMISSIONS

So how might we avoid the tragedy of the commons discussed in Part Two? Using the same example, how can we keep too many cows from grazing on the pasture? Elected officials responsible for overseeing the public lands would have two choices: they could cap the number of cows allowed on the pasture, or they could apply a tax to each head of cattle. In the first case, an auction would need to be held. Farmers would bid on the rights to have a cow graze on the land. Later these same farmers could sell or lease their rights to other farmers if they wished. In fact, this is how many municipalities dole out taxicab permits. A certain number of permits are auctioned off. Owners can then sell or lease these permits to other taxi drivers.

Both methods of limiting cattle on the field have the same net effect. They put a price on the use of the public lands. If too many cows are on the pasture, officials can either reduce the allowable cap or increase the tax. Farmers would prefer the tax approach as it gives them financial certainty. They know exactly how much it will cost to have a cow feed off the land. The environment prefers the cap method since it directly limits the number of cows permitted on the pasture. It's easier to control the total number of cows by direct regulation.

We can translate this example to the global warming problem. The atmosphere has traditionally been viewed as an unregulated dumping ground. There is no cost associated with emitting greenhouse gases. Economists call this a **market failure**. To correct this failure a price is needed on emissions. This allows individuals and businesses to find the most cost-effective means of reducing their own emissions. I have yet to meet an economist who disagrees. The debate comes down to how the price should be added. As in the example above, there are two approaches. These are through a cap or a tax. Some like to use the words *levy* or *fee* instead of *tax* since most members of the general public view the "t" word negatively.

A cap or fee can be applied in one of two places. An upstream cap or fee is applied when fossil fuels are taken out of the ground and passed to a distributor (or when the fuel is imported). A downstream cap or fee is applied when the fuel is burned and the emissions are released. Here a fee would be applied at the pump in a gas station. A cap would be applied on emissions from individual factories. When a cap is imposed, a fixed number of permits are auctioned off. These permits specify the total carbon content of fossil fuels to be supplied or greenhouse gases to be emitted. The number of permits available decreases with time, so the price for a permit will likely go up. This allows certainty in the overall quantity of emissions being reduced. When a fee is imposed, a fixed price is added to the cost of each fossil fuel. The fee increases with time. This allows certainty in the overall cost of emission reductions. In all cases, price increases are passed along to the consumer.

Permits acquired under a cap program are tradable. By allowing trading, the market finds the most efficient way to reduce emissions or the carbon content of fuels. For example, let's suppose a number of factories are each emitting 100,000 metric tonnes of carbon dioxide per year. Under a downstream cap program, each might be required to reduce its annual emissions by 10,000 metric tonnes over the next decade. Some factories would find it easy to reduce emissions. Others would find it more difficult. A few factories might reduce emissions by more than 10,000 metric tonnes. They would be allowed to sell their surplus reductions to those factories that were finding it more costly to do so. In theory, lowest-cost reductions are then achieved first. I've just outlined what have become commonly known as **cap and trade** programs.

Governments generate revenue when either a cap or a fee is introduced. The question remains as to what governments should do with this money. Economists argue that making the cap or fee **revenue neutral** is important so as not to overly burden the economy. This can be achieved by reducing taxes elsewhere. For example, income and corporate taxes could be reduced. Others have argued that the revenue generated could be returned as an annual *dividend* to all taxpayers. In both cases, consumers would have more money in their pocket. If they chose to spend it on less-fossil-fuel-intensive products, they would end up being ahead. Finally, others have suggested that governments should use some of the revenue to invest in green infrastructure. This might include improving public transit.

So which approach is the best? Economists agree that upstream pricing is the easiest way to ensure that the entire economy is affected fairly. Critics of cap and trade schemes point out that they are difficult and costly to regulate. They take a long time to set up and are open to abuse by special interests. And these schemes are less transparent to consumers. Critics of the carbon tax or fee argue that it might unfairly penalize lower- and middle-income people. But this can be alleviated by the way money is redistributed to ensure revenue neutrality. Critics also note that fees don't control the quantity of emissions directly. Businesses overwhelmingly favor the carbon fee as it gives them price certainty and is less costly for them to implement.

Different people have different opinions on how a government should use the revenues it receives from either a cap or fee pricing scheme. A diversity of views will always exist. One thing is certain. Until we put a price on greenhouse gas emissions, the tragedy of the commons tells us that we will not solve the global warming problem. The actual details of the exact pricing scheme to be used are not of primary importance.

We have more than two decades of evidence (see Figure 1.3) to show that relying on voluntary greenhouse gas emission reductions simply does not work.

THE UNITED NATIONS FRAMEWORK CONVENTION ON CLIMATE CHANGE

An international treaty to combat global warming was put together in 1992. Canada, the United States and another 191 nations are all parties to what is called the United Nations Framework Convention on Climate Change (UNFCCC). The objective of the UNFCCC is the "stabilization of greenhouse gas concentrations in the atmosphere at a level that would prevent dangerous [human] interference with the climate system." While the UNFCCC does not define what "dangerous" means, it gives some guidance. The treaty states that steps should be taken to ensure that:

1) ecosystems can adapt naturally;
2) food production is not threatened;
3) economic development can proceed in a sustainable manner.

Obviously it is impossible to come up with a definition of "dangerous" that is acceptable to everyone. Two meters of sea-level rise may be devastating to New Orleans, but it's not going to affect a person living in Denver. Citizens of Seattle might be excited about the prospects of more frequent summer heat waves. Those living in Phoenix might be concerned. And those living in Niger might be extremely worried. Nevertheless, in recent years a broad international consensus seems to have emerged: global warming should be kept to less than 3.6°F (2.0°C) above preindustrial levels. That's 2.3°F (1.3°C) further warming from today. In fact, the nonbinding United Nations Copenhagen Accord that 114 countries took note of in December 2009 states that

"deep cuts in global emissions are required...
with a view to reduce global emissions so as to
hold the increase in global temperature below
2 degrees Celsius."

MEETING THE TWO-DEGREE TARGET

Science will never be able to tell society how
much warming should be considered dangerous.
Such a decision requires value judgments to be
made. These involve social and political discussions and priorities. But science can inform
these discussions by outlining the consequences
of different levels of future warming. Science
can also calculate the quantity of allowable
greenhouse gas emissions required to keep us
below some level of future warming.

In recent years there has been a profound
advance in our understanding of how carbon
dioxide circulates within the climate system.
We know, for example, that it takes a very
long time for **natural sinks** to take up the

human-produced carbon dioxide. Suppose that we burn all known fossil fuel reserves. Then 75% of the resulting carbon dioxide will stay in the atmosphere for about 1,800 years; 25% will remain for longer than 5,000 years. We've also learned that to keep warming below any threshold requires emissions eventually going to zero. If we decide to travel down this path, it's likely that we will need to implement technologies to scrub carbon dioxide from the air. These technologies would produce *negative* emissions. They would reduce the concentration of atmospheric greenhouse gases.

The scientific community has also firmly established that the overall magnitude of global warming is determined by the total human emissions of carbon dioxide. For example, the eventual warming is the same whether 100 billion metric tonnes of carbon dioxide is emitted over ten years or over fifty years. This result also has important implications for international policy negotiations. Looking back at Table 1.1, we see that India presently emits almost three times as much carbon dioxide as Canada. It turns out that India's population is about thirty-four times as large as Canada's. However, India and Canada have emitted about the same total amount of carbon dioxide since preindustrial times. That is, both have contributed equally to the problem we have today.

Now let's suppose that we have collectively reached a global consensus that warming

	Acceptable probability of exceeding threshold	
	< 10%	< 33%
3.6°F (2.0°C)	249	1789
5.4°F (3.0°C)	1166	3623
7.2°F (4.0°C)	2083	5309

TABLE 3.1: *Allowable total emissions of carbon dioxide from January 1, 2011, onward in billions of metric tonnes. The indicated values keep warming below a given threshold with some probability. The first column defines the global warming threshold above preindustrial times. The second column shows the allowable emissions of carbon dioxide assuming we accept only a one-in-ten chance of breaking the warming threshold. The third column is the same as the second column but for a one-in-three chance.*

should be limited to 3.6°F (2.0°C). Let's further suppose that we accept only a one-in-ten chance of passing this threshold. As shown in Table 3.1, we can put 249 billion metric tonnes of carbon dioxide into the atmosphere before going to zero emissions. Present-day emissions are 37 billion metric tonnes and growing. This means that we can continue to emit at current rates for 6.7 years before immediately going to zero worldwide. Obviously this isn't going to happen. So let's accept a one-in-three chance instead. These are good betting odds in Vegas. But you wouldn't play Russian roulette with two of the six chambers loaded. You probably also wouldn't fly in an airplane if it had a one-in-three chance of crashing.

By accepting a one-in-three chance of passing the 3.6°F (2.0°C) warming threshold, we have a grand total of 1789 billion tonnes of allowable future carbon dioxide emissions to play with. That's forty-nine years at current rates. But we would have to instantly go to zero emissions thereafter. If, instead, we immediately started slowly reducing emissions, we would have something like ninety-eight years until we had to go to zero emissions. Similar calculations can be done for 5.4°F (3.0°C) or 7.2°F (4.0°C) warming levels (see Table 3.1).

THE FAILURE OF INTERNATIONAL TREATIES

The Kyoto Protocol to the UNFCCC was adopted in 1997. It was designed to put some teeth into international efforts aimed at trying to avoid dangerous human interference with the climate system. Developed countries agreed to reduce greenhouse gas emissions to 5% below 1990 levels, on average, by 2008 to 2012. Canada's target was 6% below 1990 levels. The US target was 7% below 1990 levels. While both Canada and the United States signed the protocol, it was never ratified in the United States. Canada ratified the protocol on December 17, 2002. By 2008, Canada's emissions were 24%

above 1990 levels. US emissions had increased by 14%. On the other hand, UK emissions had decreased by 18% and Germany's by 22%.

In Europe the Kyoto commitments were taken very seriously. This was not the case in either Canada or the United States. In fact, the Canadian government eventually announced that it would not try to meet its Kyoto target. Instead, in 2007, a Made-in-Canada solution was proposed. Now Canada argued that it would reduce its emissions to 3% below 1990 levels by 2020. Three years later, Canada changed its tune yet again. This time it was going to match the US number proposed by President Barack Obama as part of the Copenhagen Accord. Obama promised the United States would reduce emissions by 17% relative to 2005 levels by 2020. This was equivalent to saying US emissions would reduce 3.6% from 1990 levels by 2020. In Canada's case, this meant emissions would actually rise 2.5% from 1990 levels by 2020. Are you confused? Imagine what the rest of the world is thinking.

A quick look back at Figure 1.3 shows that the UNFCCC and its Kyoto Protocol have not been able to slow the growth in greenhouse gas emissions. The Copenhagen Accord, designed as the successor to the Kyoto Protocol, is also a monumental failure. There is a staggering disconnect between international policy negotiations and science. The voluntary reduction targets submitted by nations as part of the Copenhagen Accord will guarantee that the 3.6°F (2.0°C) warming threshold will be surpassed. Yet the Copenhagen Accord was specifically put together to ensure this was not the case. Even if all countries abide by their Copenhagen Accord targets, it is reasonably safe to assume that 5.4°F (3.0°C) and perhaps even 7.2°F (4.0°C) warming will occur over the course of the next century or so.

It seems clear that the international treaties to date have been a failure. The negotiations play out like a textbook example of the tragedy of the commons. Why should I reduce emissions? I bear all the costs of doing so. The costs of inaction are distributed among everyone. Perhaps a more useful role for the UN at this juncture is to try to reach agreement on an internationally acceptable price on carbon emissions. Alternatively, if 3.6°F (2.0°C) warming is to be avoided, perhaps negotiations could focus on how to allocate the 1789 billion metric tonnes of allowable future carbon dioxide emissions. If society wants to deal with global warming, we will likely need to start with grassroots initiatives. We are already beginning to witness signs of this occurring in schools, businesses and communities all across North America.

GET OUT AND VOTE

Given the failure of our national political leaders to put in place measures to deal with global warming, what can an individual do? For a start, we can vote. In Canada, voter turnout has been declining steadily over the last fifty years (see Figure 3.1). In the 2008 general election,

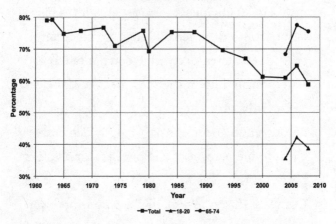

Canadian Voter Turnout

FIGURE 3.1: *Voter turnout since 1962 in Canadian federal elections (above) and US presidential elections (below). The squares indicate overall turnout. The triangles indicate youth turnout (between 18 and 20 in Canada; between 18 and 24 in the US). The circles indicate turnout by seniors (between 65 and 74 in Canada; older than 65 in the US). Source: Elections Canada and US Census Bureau.*

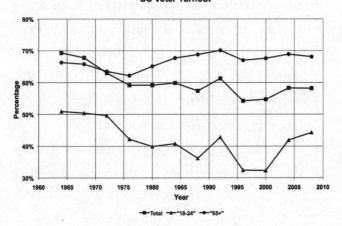

US Voter Turnout

only 59% of eligible voters cast a ballot. This is down from 79% in 1962. Things are no better in the United States. Sixty-nine percent of eligible voters turned up in the 1964 presidential election. Only 58% voted in 2008. Voter turnout among young people is particularly deplorable. Thirty-nine percent of eligible Canadian voters between the ages of 18 and 20 voted in 2008. Compare this to 76% of eligible voters between the ages of 65 and 74 who turned up at the polls. Youth voting in 2000 reached an all-time low in the United States. Only 32% between the ages of 18 and 24 bothered to cast a ballot. Two out of every three young people stayed at home. Even with the power of social media such as Facebook and Twitter, voter turnout among youth in the United States only reached 44% in the last presidential election. Sixty-eight percent of seniors voted at that time.

Global warming is going to manifest itself on the next generation. Yet that generation is not showing up at the ballot box. The youth of today may be able to blame their elders for the predicament that we currently face, but they can only blame themselves for not getting involved in electing leaders who will deal with it.

Municipal elections are just as important as state, provincial and national ones. Mayors and councilors determine zoning bylaws, park designation and transportation corridors. They determine how our property taxes are spent on the operation and design of our communities. If we plan to start reducing our dependence on fossil fuels, we will need to elect leaders who possess and can articulate a vision for a sustainable future.

FIGHTING *FOR* CHANGE

As explained in the previous section, if we really want to solve the global warming problem, the first step is to vote. This is especially true for the youth of today. On the way to the polls, collect five of your friends and get them to vote as well. Or chat with your grandparents and try to persuade them to vote for your generation. But voting is not enough. Once political leaders are elected, they must be supported in their efforts to facilitate change.

It's relatively easy to mobilize large groups of people against a particular initiative. Let's suppose a municipality wants to install a few windmills on a local hilltop. Local council might want to produce enough electricity to power their municipal operations. You can easily imagine groups of residents protesting that their view will be impaired. Letters will be written to local newspapers arguing against the project. Commercials might start appearing on TV from newly formed groups with names like Citizens Against Windmills on the Hill. But where will support for the windmill project come from?

We need to produce electricity somehow. While energy conservation is extremely important, it alone cannot possibly take us to zero emissions. We need to electrify our transportation systems. Instead of driving noisy and polluting automobiles, we'll have to drive electric cars. These cars need access to power to charge their batteries. In the future, electricity demand will go up even with stringent conservation measures. And no matter how we produce electricity, there will be an environmental footprint. It's important to recognize this and try to ensure that the footprint is as small as possible. Environmental organizations need to switch their tactics. They are experts at fighting against particular projects. Now they will need to learn how to advocate in support of renewable energy projects.

As I have mentioned, most of the renewable energy systems are readily available. They just require a price to be put on greenhouse gas emissions to fix a market failure. This will only occur if we elect politicians who will implement carbon pricing and if we support them in their efforts to do so. Some continue to argue that industry should be allowed to reduce emissions through voluntary measures. If you look back at Figure 1.3, you'll see that several decades of voluntary reductions have not worked. Until the market failure is corrected, it's unlikely we will see reductions anytime soon.

REPLACING ONE PROBLEM WITH ANOTHER

In recent years the controversial idea of **geoengineering** has been suggested as a potential technological fix for global warming. In geoengineering, humans deliberately alter the climate in an attempt to counterbalance the warming from increasing greenhouse gases. Several outlandish possibilities have been bandied about. One involves continuously releasing tiny particles in the upper atmosphere to scatter sunlight back to space. This would have the effect of cooling the Earth. Another idea is to put satellites in space with retractable mirrors that block some sunlight. The list goes on: promote the formation of more low clouds; put reflective blankets over ice sheets; sprinkle iron in the ocean and hope that more phytoplankton growth will draw down carbon dioxide.

The bottom line in all this is that we have no idea whether or not these activities will have the desired effect. We also don't know what other problems would be created as a consequence of their implementation. Let's suppose that you are addicted to smoking. Medical experts around the world are united. They tell you that smoking is bad for you. The chance of getting lung cancer is much increased. What's the best thing to do? Quit smoking? Or wait until you get cancer and then start taking drugs in the hope that one of them will yield a cure? The answer is pretty obvious. It's the same with global warming. We know what the cause is. The smart cure is to eliminate the source of the problem in the first place. This means reducing emissions.

A CHANGE IN BEHAVIOR

Technology itself will not solve global warming. Individual behavior and consumption patterns will need to change as well. For too long we have lived by the axiom that growth is great. We strive for economic growth year after year. We drive it by increasing population. But infinite growth cannot occur in a finite system. Collapse is inevitable.

Consumers have the power to change the supply of goods and services by changing their demands. Take organic foods, for example. Most grocery stores now proudly advertise their organic fruit and vegetable sections. But this is a relatively recent phenomenon. The whole organic food industry popped up in response to increasing consumer demand for pesticide-free products. These same consumers can also demand electricity-sipping cars instead of gas-guzzling vehicles. They can buy energy-efficient products and boycott overly packaged items. They can start demanding every product be tagged with its carbon footprint. The footprint lets you know how many greenhouse gases were produced in manufacturing the product and delivering it to the store. Consumers can choose which products to purchase based on their effect on the environment. Products produced abroad using coal-powered factories don't stack up as well as locally manufactured goods using renewable energy resources. Consumers can also insist that companies take responsibility for end-of-life recycling of their products. This would aid in eliminating built-in obsolescence that promotes wasteful consumerism.

It takes the concerted effort of only a few to begin new trends. As more and more people start to switch their behavior, we get closer to a tipping point. Beyond that critical moment, a transitional avalanche occurs. This can apply to the affordability of new technologies as well as the acceptance of new ideas. Take recycling as an example. Today almost every household recycles paper, cardboard, glass, metal and plastics through curbside pickup programs. This wasn't always the case. While recycling in one form or another has been around for centuries, widespread recycling didn't start until the 1970s. Escalating energy prices and education played a crucial role in turning recycling into a social norm.

We are at a pivotal juncture in human history. Two paths lead ahead. One of them takes us to the tragedy of the commons. The other leads us to a sustainable livelihood. In this case we strive for equilibrium. So what can the individual do to help? Switching from incandescent to fluorescent lightbulbs helps conserve energy. Employing reusable instead of plastic shopping bags cuts down on waste.

But these efforts are not enough. They are baby steps down the sustainability path. A much more fundamental shift in our behavior is needed. This will require a reexamination of what we believe is important to us.

How do you currently feel when you see someone driving a Hummer? Do you say to yourself: *Wow, nice car*? Or do you think: *How selfish of you to drive that gas-guzzling vehicle*? Most of us probably would think the former. How often have you sat in your car while it idles for minutes on end? Wasteful use of energy will need to become as socially unacceptable as smoking in public spaces. The same is true for using the atmosphere as an unregulated dumping ground for our greenhouse gas emissions. We will need to continually ask ourselves how our individual actions are affecting the livelihood of others. If we want to solve global warming, the *Me Generation* must evolve into the *Us Generation*.

Embracing Change

Scientists now have a remarkably good understanding of the causes and consequences of global warming. Atmospheric scientists know that increasing greenhouse gases cause global warming. They know this as surely as medical professionals understand that smoking causes cancer. We stop smoking if we want to lower our risk of developing lung cancer. We must stop emitting greenhouse gases to the atmosphere if we want to stop global warming.

Of course, there is, and always will be, scientific uncertainty. But this uncertainty is about how global warming will manifest itself on the local scale. It's not about whether or not glaciers, permafrost or sea ice will melt. Nor is it about whether or not sea levels will rise. Or whether there will be an increased likelihood of extreme rain events and summer droughts. Scientific uncertainty is not a justification for inaction.

Based on all the evidence before us, the only compelling argument for inaction is that as a society we do not believe that we have any responsibility for the well-being of future generations. But of course our actions suggest otherwise. It's a basic human instinct to want to provide for our children and to offer them a better life. We've probably written wills to ensure they are looked after in case something happens to us. On Memorial Day in the United States and Remembrance Day in Canada we commemorate those who gave their lives to ensure that we could live better and more peaceful lives. We have created national parks and nature reserves to conserve our ecosystems for future generations. We have banned chemicals that deplete the ozone layer and mandated the removal of lead from gasoline. We have set standards to reduce harmful chemical emissions that cause acid rain. Forest companies must replant trees, and mining companies must reclaim the landscapes, once the resources have been extracted. We have taken steps to ensure that we don't pollute our streams, lakes and oceans. Why? To preserve them for future generations. So what is the problem? Why is it that so many people apparently reject the evidence in front of them?

Vested interests have spent an inordinate amount of time and money trying to convince the public that the science of global warming is either uncertain or that the whole issue is somehow a hoax. Some of these vested interests worry about the effects of potential solutions on their bottom line. They might believe that it's in the interest of their shareholders to delay government action until they've positioned their company to thrive in a low-carbon economy. There are also those in society who oppose any and all forms of government regulation. These are the so-called libertarians. Others fear the formation of an uber-governing body in Geneva under the auspices of the UN that might somehow limit the autonomy of Canada or the United States.

There is no question that doubt about climate science is on the rise because of a well-orchestrated campaign of misinformation. But this doesn't explain the lack of action by those who understand the significance of the science. Why is this the case?

Facing the challenge posed by global warming requires each of us to confront our own high-carbon lifestyles. This is something that doesn't come naturally. Our instinct is to distance ourselves from the problem. Perhaps we find it disturbing and so focus on other more pressing issues in our daily lives.

We might believe that we've done our part already. It's up to others now. Or we might avoid the problem altogether out of a sense of hopelessness—that feeling of *What can I do to make a difference anyway?* Our inability to confront our own roles makes it much easier for us to selectively attach ourselves to messages that we *want* to hear. We take solace in the belief that somehow the science is uncertain, that corrupt scientists have conjured up a hoax. It's a natural self-defense mechanism. But deep down we may have our doubts.

Global warming has been branded an environmental problem. But it is really an economic and social problem. We've spent too much time living within a culture of global-warming fear and denial. It's time to recognize global warming for what it is: the most self-empowering issue we will ever face. Every consumer of energy is part of the problem. Every person is therefore part of the solution. We are entering an age of creativity and innovation unlike any that modern society has experienced before. Rather than fearing this change, we need to embrace it. And the change starts in each and every one of our households. The time for *us* is now.

Glossary of Terms

biosphere—all the world's living organisms and their ecosystems

cap and trade—a type of policy that puts a price on carbon dioxide emissions. The "cap" sets the limit as to how much carbon dioxide can be emitted. Carbon dioxide emitters can "trade" their emission permits amongst themselves.

climate—the statistics of weather. This includes average weather and the likelihood of occurrence of a particular weather event. The United Nations World Meteorological Organization defines *normal* climate as weather conditions averaged over a thirty-year period.

dieback—when applied to forests, dieback means the death of trees and subsequent retreat of forest cover

electromagnetic radiation—energy in the form of waves that travel at the speed of light. Electromagnetic radiation is all around us. Radio waves, microwaves, infrared radiation, visible light, ultraviolet radiation, X-rays and gamma rays are all forms of electromagnetic radiation.

experimental—based on an experiment. An experiment is a procedure that is designed to test a hypothesis. Experiments are performed under controlled conditions.

fossil fuel—a fuel (such as coal, oil or natural gas) that was formed many millions of years ago from the remains of prehistoric plants and animals

frequency—the number of wave crests that pass a given point in a second

geoengineering—deliberately altering the climate in an attempt to counterbalance the warming from increasing greenhouse gases

global radiative equilibrium—the state when the total amount of energy that the Earth receives from the sun is equal to the total amount it emits back to space

global warming—the average warming of the Earth's surface temperature as a consequence of human activity

greenhouse effect—process by which infrared radiation is absorbed by greenhouse gases and re-emitted back to the Earth's surface, thereby preventing heat from escaping the Earth system

greenhouse gas—certain atmospheric gases such as water vapor, carbon dioxide and methane that are effective at absorbing infrared radiation from the Earth

hypothesis—an educated guess or explanation of how a natural process operates

industrialization—the process of transforming rural, agricultural societies into industrial ones

market failure—a term used in economics when the free market is not efficient in the allocation of goods and services

natural sink—a naturally occurring process or reservoir that can take up and store a greenhouse gas for an extended period of time

photosynthesis—process in plants by which the sun's energy is transformed and stored as chemical energy by combining with carbon dioxide from the atmosphere and water to form sugar; oxygen is then released back to the atmosphere in the process

positive feedback—occurs when a change in one process leads to a change in a different process that in turn makes the change in the original process larger

precautionary principle—requires that, prior to doing something that could potentially cause harm, one must show that the procedure will not cause a problem instead of proving that a problem will exist if the procedure is followed. This takes the burden of proof off those who might suffer from an action and puts it on those who would undertake the action.

prediction—a forecast made on the basis of scientific understanding of how a process works

radiation—*See* electromagnetic radiation

revenue neutral—a procedure that does not create any net revenue for the government. A revenue neutral tax is one that offsets any revenue gained from the introduction of the tax by tax reductions in other areas of the economy.

scientific method—process used to study the natural world and human society; it is comprised of four stages: observational, hypothesis, prediction and experimental

tipping point—critical moment in time after which a cascade of change occurs

trace gases—1% of the atmospheric composition, including very small amounts of water vapor, carbon dioxide, ozone, nitrous oxide and methane

wavelength—the distance between the crests of two waves

weather—the state of the atmosphere at a particular place and time described in terms of temperature, cloudiness, windiness, precipitation, etc.

DR. ANDREW J. WEAVER is professor and Canada Research Chair in climate modeling and analysis in the School of Earth and Ocean Sciences, University of Victoria. He was a lead author in the United Nations Intergovernmental Panel on Climate Change, co-recipient of the Nobel Peace Prize in 2007. He is also the author of *Keeping Our Cool: Canada in a Warming World* (2008).